Visit the World-renowned

ORANG-UTAN REHABILITATION CENTRE at Sepilok

and relax in Paradise…

Sepilok Palm Resort

(located next to the Centre)

Sandakan office:
1st Fl., Lot 7, Block A, Bandar Pasaraya,
Mile 4, North Road, PPM 255 Elopura,
90000 Sandakan, Sabah, Malaysia.
Tel: 6089-228081 Fax: 6089-271777
e-mail: sepilok@po.jaring.my

Tawau office:
1st Fl., No. 484, Block P, Bandar Sabindo,
P.O. Box 61120, 91021 Tawau,
Sabah, Malaysia.
Tel: 6089-765200 Fax: 6089-763575, 763563
e-mail: psrt@po.jaring.my

Visit our web site at: http://www.sipadan-resort.com

The Natural History of
ORANG-UTAN

The Natural History of
ORANG-UTAN

Elizabeth L. Bennett

with photographs by
Tengku D.Z. Adlin, C.L. Chan, Arthur Y.C. Chung,
Phillip Cribb, Hans P. Hazebroek, Ku Min Yong, Elkie Lies,
Therese Lillieskold, Aaron Law, Jimmy Omar, Stephen Von Peltz,
Andrew Plumptre, W.M. Poon, Tan Swee Heng,
Albert Teo, Tham Nyip Shen, William Wong,
Sylvia Yorath, and the author

Natural History Publications
Kota Kinabalu

1998

Published by

Natural History Publications (Borneo) Sdn. Bhd., (Company No. 216807-X)
A913, 9th Floor, Wisma Merdeka,
P.O. Box 13908,
88846 Kota Kinabalu, Sabah, Malaysia.
Tel: 6088-233098 Fax: 6088-240768
e-mail: chewlun@tm.net.my

Half title page: Adult male Bornean orang-utans develop impressive broad face flanges.
                     (*Photo: Jimmy Omar*)
Frontis piece: "Orang Utan attacked by Dyaks".
                     Reproduced from A.R. Wallace, *The Malay Archipelago.*

First published 1998.

The Natural History of Orang-utan
    by Elizabeth L. Bennett

Design and layout
    by C.L. Chan

Perpustakaan Negara Malaysia Cataloguing-in-Publication Data

Bennett, Elizabeth L.
    The natural history of orangutan / Elizabeth L. Bennett ;
    with photographs by Tengku D.Z. Adlin ... [et. al.].
    ISBN 983-812-121-9
    1. Orangutan. 2. Wildlife conservation. I. Tengku D.Z.
    Adlin. II. Title.
    639.979883

Printed in Malaysia

To

My parents

Contents

W.M. Poon

Foreword

Throughout the world, orang-utans have long been the most famous of Borneo's animals. Closer to home, myths and legends of orang-utans are woven deeply into the rich cultural fabric of many of our peoples. In recent times, the orang-utan has been adopted as Malaysia's mascot, and as the symbol of the 1998 Commonwealth Games being held in Malaysia. Orang-utans are also a major part of the tourism draw to the various political regions of Borneo. Yet in spite of their importance, and their prominence in our minds and our culture, orang-utans are becoming increasingly endangered due to human activities. Clearance of their forest home, and hunting of the animals for food, trophies and the pet trade have reduced their range and their numbers dramatically in recent years. Although some loss of forest is inevitable as Malaysia and Indonesia develop, equally essential is that we guarantee that these magnificent animals have a secure natural home, in the wild. This means that we must set aside large reserves, and protect those reserves effectively on the ground. Only then can we realistically hope that orang-utans, and all of the other countless forms of life in our forests, will survive to enrich future generations. A crucial element of this process is garnering public support for, and appreciation of, the orang-utan. This extremely attractive book, full of superb photographs as well as being highly informative, is an invaluable contribution to promoting this awareness, and hopefully in generating enthusiasm for conserving this most magnificent creature. I congratulate the author, all of the photographers, and Natural History Publications in producing such an excellent book. It will certainly enhance other efforts being made by numerous people and agencies throughout Borneo and Sumatra, to try to ensure that the orang-utan has a bright future in these great forests of ours.

Yang Berhormat Tan Sri Bernard G. Dompok
Minister of Tourism and Environmental Development, Sabah

P.J. Cribb

Introduction

D awn in the rain forest of Borneo. Mist hangs in the air between the branches and lianas all around and rising high above. The cicadas, frogs and early birds call, unseen through the leaves in the dim early morning light. A gibbon starts its bubbling, haunting song across the hill in the next valley. Suddenly, in the trees nearby, there is a gentle crash, and some leaves start to move. A few minutes later, a hairy red arm can be seen, followed by a face with huge brown eyes staring down. It is a juvenile Bornean orang-utan, moving around for his first feed of the day. A louder crash in the background is accompanied by a noisy kissing sound; his mother comes briefly into view, to check if her son is in danger. The large adult female has a pair of tiny hands clinging to her side, as her latest offspring peers through her mother's long hair to peek at what's going on. After a few minutes, the female swings slowly off, and disappears into the foliage and mist, followed by her more agile young son.

This has been a rare glimpse of the least known, and most enigmatic of the great apes: the orang-utan. Of our closest relatives, the great apes, the orang-utan is unique in many ways: it is the only great ape in Asia— indeed, the only one outside Africa. It is the most arboreal and least sociable of all the great apes. Yet it is also under threat. It only occurs in limited parts of the islands of Borneo and Sumatra, where its habitat is disappearing to make way for oil palm plantations and other forms of development, and it is hunted in much of its range.

In this small book, we take a glimpse into the life of this beautiful red ape: where it lives, what it eats in the wild, how it moves around the forest, and its unusual social life. Then we consider the orang-utan's conservation and future: what are the main threats facing it, how threatening are they, and what is being done about them? The book concludes with a short section on where and how visitors can see orang-utans in the forests of Sumatra and Borneo.

(opposite) The orang-utan is the most enigmatic and least known of the great apes.

1

Tan Swee Heng

1
Myths, Legends and First Impressions

O rang-utans, like other great apes, have long been a source of major fascination for mankind. This is undoubtedly due to their often being uncannily like humans and highly intelligent, at the same time as being large and extremely strong.

As a result of this fascination, all of the peoples of Sumatra and Borneo who share their forests with orang-utans have developed myths, beliefs and legends about these red apes. Some believe that humans descended from orang-utans, and others that humans turn into orang-utans when we

S. Von Peltz

(above) Orang-utans have long been a source of considerable fascination for mankind.
(opposite) Orang-utans are large and powerful, yet uncannily human-like in their expressions.

3

C.L. Chan

(above and right) The apparently knowing expressions and intelligence of orang-utans have fostered many myths, legends and beliefs about the species and its close links to humans.

die. Tales are sometimes told of wild orang-utans attacking or raping humans, and of humans and orang-utans marrying and producing half-man, half-ape children. In Borneo, if a woman has an ugly baby, this might be ascribed to her having looked at an orang-utan during her pregnancy or immediately after the child was born. For some people in Borneo, it is considered bad luck to look into the face of an orang-utan, and laughing at one can bring misfortune to the whole longhouse community.

W.M. Poon

On the other hand, in some of the stories the orang-utan is considered highly benevolent. Tales are told of how orang-utans saved an ancestor or other relative from disaster. The Ibans in parts of Sarawak recount how orang-utans helped them in the past. Long ago, the story goes, humans used to give birth by opening up an expectant mother with a machete. Then one day, a woman was about to give birth in the forest when a female orang-utan saw her. The orang-utan carried the woman up into her tree-top nest, and showed her how to give birth properly. Even today, this and similar beliefs protect the orang-utan from being hunted or otherwise harmed by Ibans in some areas.

Western explorers were equally fascinated by the first glimpses and descriptions of orang-utans—although with some of the very earliest descriptions, it was unclear if they were describing orang-utans, or some of the local people whom they met. For example, a Dutch doctor Jacob de Bondt is credited with the first ever written description of wild orang-utans, but his portrait of a female who seemed to have an idea of modesty, covering herself with her hand on the

Albert Teo

appearance of men with whom she was not acquainted has led people to question whether this was indeed an orang-utan or an unusually hairy woman! In 1712, Captain Daniel Beeckman gave an authentic description of orang-utans in southern Borneo, and wrote: "The Monkeys, Apes and Baboons are of many different Sorts and Shapes; but the most remarkable are those they call Oran-ootans, which in their language signifies Men of the Woods: these grow to be up to six Foot high; they walk upright, have longer arms than Men, tolerable good Faces (handsomer I am sure than some Hottentots that I have seen), large Teeth, no Tails nor Hair, but on those parts where it grows on human bodies."

By the time naturalist Alfred Russel Wallace undertook his famous explorations of the Malay Archipelago in the mid-1850's, written descriptions of orang-utans were becoming much more accurate. For example, Wallace said: "The orang-utan is known to inhabit Sumatra and Borneo, and there is every reason to believe that it is confined to these two great islands... The long and powerful arms are of the greatest use to the animal, enabling it to climb easily to the loftiest trees, to seize fruits and young leaves from slender boughs which will not bear its weight, and to gather leaves and branches with which to form its nest... Each mais (orang-utan) is said to make a fresh one for itself every night."

Even so, much of its behaviour was to remain unknown for many more years. In 1916, Robert W. Shelford commented that "considering its size, the Maias (orang-utan) is remarkably inconspicuous in its natural surroundings. Until men can acquire arboreal habits it seems likely that the domestic arrangements of the ape will remain undis-covered." Indeed, virtually all of the descriptions of the late-nineteenth and early-twentieth centuries were written

(left) Their shyness, and the fact that they spend their time in trees in dense tropical forests meant that for many years, the lives of orang-utans in the wild were largely unknown.

by people whose main intent was to collect specimens for museums, and their writings were more about catching and killing the animals than about their behaviour. Wallace himself describes killing 17 orang-utans in his expedition of 1855, and twenty years later, William Hornaday collected an incredible 43 animals from western Sarawak for American natural history museums. Such killings for museums peppered the writings of naturalists right through to the 1930s.

To appreciate why so little was known of orang-utans for so many years, consider this. Most orang-utans live in places which are either horribly steep, or deeply swampy. The animal spends most of its time high in the dense trees of a tropical rain forest, on its own, and is a shy creature. So all of these things make it extremely difficult to study in the wild. This meant that our knowledge of wild orang-utans was a mixture of truth, myths and legends until very recently. It was not until George Schaller's surveys in 1960, and subsequent detailed field studies by David Horr, Biruté Galdikas, John MacKinnon, Herman Rijksen, Peter Rodman and others from the 1970s onwards that we have obtained a more accurate picture of the natural behaviour of the orang-utan.

W.M. Poon

Originally, the orang-utan was named after "a sylvian deity given to merriment and insatiable lasciviousness".

2
What are Orang-utans?

Orang-utans belong to the order Primates, which is the large group of animals comprising lemurs and lorises, monkeys, apes, and humans. So the primates include ourselves and all of our closest relatives.

Within the Primates, orang-utans are classified as great apes, together with the common chimpanzee, pygmy chimpanzee and gorilla. Great apes are distinguished from monkeys by their greater brain size and intelligence, and by their not having a tail.

For many years, there was a hot debate amongst biologists about which of the four great apes was our own closest relative. Anatomically, it seemed to be the chimpanzee, but in the 1980s, some scientists made a

The only other Asian apes are the gibbons, or lesser apes. Orang-utans share their forests with four species of gibbons: the Bornean gibbon in most of Borneo, agile gibbon in south-west Borneo, and lar gibbon and siamang in north Sumatra.

E.L. Bennett

(right) A mountain gorilla from central Africa. Together with orang-utans and the two species of chimpanzee, these great apes are our closest relatives.

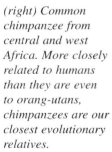

(right) Common chimpanzee from central and west Africa. More closely related to humans than they are even to orang-utans, chimpanzees are our closest evolutionary relatives.

A. Plumptre

Elke Lies

*(above) Sumatran orang-utan
often sport luxuriant beards.*

*(left) Adolescent male Bornean
orang-utan.*

strong case for the orang-utan winning the contest. These arguments were based as much on which of the great apes people studied and felt particular affinities for, as on scientific evidence which was confusing at best. In the 1990s, genetics came to the rescue and solved the problem once and for all. It is now clear that chimpanzees are indeed our closest living relatives—and to people's amazement (and sometimes horror), it was found that we share a staggering 98.4% of our genes with the pygmy chimpanzee! Of the great apes, orang-utans are furthest from us, and also from each of the other great apes. So we are more similar to chimps and gorillas, than gorillas and chimps are to orang-utans. Even so, 96.4% of our genes are identical to those of orang-utans, so they are still very close relations, and the historical split between the orang-utan line and the human-gorilla-chimp line only occurred some 12 to 16 million years ago.

Genetic work in the 1990s then made another amazing discovery. Ever since the orang-utan was first described more than two hundred years ago, there was thought to be one species living today, with two different sub-species or races: the Sumatran and the Bornean. But genetic studies have shown that these races are so different, we should regard them as separate species. The genetic difference between them is as great or greater than between the two species of chimpanzee, and is five to ten times greater than the distance between other sub-species of animals such as different forms of lions or of tigers. So it seems that we have not one, but two, separate species of orang-utan alive today: the Sumatran and the Bornean.

When a species is first described by science, it is given two names: the first or generic name, and the second or specific name. A generic name can be shared by several different species, but the specific name is unique to each individual species. When the orang-utan was first described scientifically in 1766, it was named *Simia satyrus*. *Simia* means ape, and *satyrus* was "a sylvian deity given to merriment and insatiable lasciviousness". This delightful, if somewhat inaccurate, image vanished early this century when the species was re-named *Pongo pygmaeus*. The newer name was even more misleading though, since *Pongo* came from a west African word *mpongwe*, referring to their local ape (probably a gorilla), and *pygmaeus* means pygmy or dwarf-like. Since the orang-utan does not occur anywhere remotely near west

Adult female Sumatran orang-utan.

Adult male Sumatran orang-utan.

Elke Lies

Africa, and is one of the largest animals in Borneo and Sumatra, the name is hardly apt. Nonetheless, it has stuck, and is the name by which the orang-utan is known scientifically to this day.

If we accept that the orang-utan is two species, not one, then the Bornean orang-utan goes by the original name of *Pongo pygmaeus*, and the Sumatran by *Pongo abelii*, named after C. Abel who first described "an ourang-outang of remarkable height" found on the island of Sumatra in 1825.

The orang-utan seems blighted by problems with its name, because its common name has also caused confusion. Orang-utan comes from orang hutan in the Malay and Indonesian languages, meaning forest man. Contrary to popular belief, however, this is not the local name for the animal, but one adopted in English by early travellers. Within its own range, the red ape is known by a variety of names, such as mawas among the Malays and in northern Sumatra, maias among the Bidayuh and Ibans, kahui by the Muruts, and kogiu or kisau by the Kadazans and Orang Sungai.

3
What do Orang-utans look like?

The orang-utan is in the record book for several reasons. It is the largest and most intelligent primate in Asia (excluding humans), and is the largest tree-dwelling animal in the world. Among primates, it is highly unusual in having long, straight, course hair, rather than the shorter, smoother coats typical of most monkeys and all other great apes. Orang-utans are amazingly powerful for their size, which gives a feeling of awe and slight alarm when watching an adult tear open a whole coconut or durian with its bare hands.

Orang-utan's feet have opposable toes which effectively give them four hands, all equally able to hold on to branches and other objects.

15

The orang-utan is the largest tree-dwelling animal in the world.

C.L. Chan

(above) Orang-utans are exceptionally agile in the forest...

(left) partly due to their having highly mobile hip joints.

S. Yorath

In captivity, orang-utans tend to become overweight, and also to grow unusually long hair. This is due to a combination of over-feeding and lack of exercise, including from spending too much time on the ground.

All orang-utans are covered with long, straight reddish hair, which has earned them the nickname of red ape. The actual shade of red varies between individuals, and also with age, and can go from a fairly bright orange in juveniles, to almost black in some very old animals. The skin of the face is largely hairless, and is a very dark bluish-black in adults. Infants start life with pale pink spectacles round their eyes and mouth, though they disappear with age as the skin darkens. Infants generally have fairly sparse fur on their heads, lower arms and legs, and the scanty, long hair on their heads often sticks up all round, to look like a somewhat unkempt golden halo. Orang-utans also have close-set, forward-facing eyes, and although their jaw is very prominent, the rest of their face is flat, so all of these features combine to contribute to their uncanny resemblance to humans.

Orang-utans have highly mobile hip joints, which allow the animals to be incredibly agile in the trees, and also opposable toes on the foot which effectively give the animals four hands: they can hold on to branches and manipulate objects almost as well with their feet as they can with their hands. Both of these features (mobile hips and opposable toes) allow orang-utans to move through trees and hang upside down from branches using just their feet to support their whole weight.

(opposite) An orang-utan's arms are longer than its total height, another feature which makes them extremely agile in the forest.

Bornean and Sumatran orang-utans differ in appearance, so an experienced observer can tell at a glance where the animal comes from. The coat colour of Sumatran orang-utans tends to be lighter, and the hair less coarse but longer. Sumatran orang-utans usually have whitish hairs on the face and groin, which Borneans never do. Sumatrans tend to be taller but slimmer than Borneans, and their faces are generally longer and thinner. The most obvious differences are between the adult males. Adult male Bornean orang-utans have larger and wider face flanges than their Sumatran counterparts, although Sumatran males make up for this by having longer beards, and often thick moustaches as well. In spite of their differences, Sumatran and Bornean orang-utans in zoos can breed with each other and, as you would expect, the offspring look pretty much intermediate between the two forms.

For both Sumatran and Bornean orang-utans, males are twice as big as females, and also look somewhat different. Adult females weigh an average of about 40 kg, whereas wild males weigh in at about 75 kg. Both sexes can become over-weight in zoos though, and obese captive males can reach a staggering 200 kg. Orang-utans in the wild rarely stand upright, but if they did, males can reach 1.5 m in height. What makes them unusual compared to upright humans is that their arm span is greater than their height—an average adult male's arm span is 2.25 m —which is even more obvious when the animals are sitting down, with their long arms curled up around their legs.

Not only are males much bigger than females, they also have features which make them appear larger still. Adult males have enormous cheek flanges either side of their face, made of fat and fibrous tissue, which look a little like the blinkers of a horse. They also have dense, longish red beards, which make their face look bigger still. Males also have a huge sac of skin on the throat which hangs down onto the chest. Inside are large air sacs which allow the animals to produce extremely loud, haunting calls which carry for up to a kilometer through the canopy of the rain forest. It has been suggested that the face flanges might help the call to carry further, by focussing it in one direction.

Unlike humans, the development of these male characteristics does not happen automatically as an orang-utan grows up. A maturing male does not develop his large cheek flaps if a dominant adult male is around. This

Albert Teo

In captivity, orang-utans sometimes stand and walk upright. They almost never do this in the wild.

has been seen both in zoos and in the wild: the development of face flanges in the young male is suppressed by the presence of an adult male, sometimes for three years or more. Take the dominant male away, and the younger male rapidly develops his cheek flanges and throat sac. The reason for this is unknown, as is the mechanism by which it works.

Another question mark lies over why male and female orang-utans are so different. In most mammals, males are larger than females if there are many more females than males in a social group. This means that there are not enough females to go round, so males have to compete strongly with other males to get those females. Under these conditions, large males tend to be more successful than small males, so genes for large males spread through the population. The greater the number of females compared to males in a social group, the greater the competition between males, so larger size becomes more of an advantage for males. But what about orang-utans? They do not live in social groups, and spend most of their lives on their own, so why are males so much bigger than the females?

The answer probably still lies in competition. Female orang-utans breed extremely slowly, and only produce one infant every five years or so. Imagine a male seeking to mate and to produce his own offspring: he finds that most females are not receptive to his advances for most of the time. So competition for the very few females who are able to conceive at any one time is great. Hence, even though the animals do not live in social groups with lots of females per male, competition amongst males for fertile females is considerable, so larger males are likely to win more females and produce more offspring than are smaller males.

Several field researchers have also suggested that sexual selection might be playing a role. This is the evolutionary process when one sex (in this case, females) choose mates with particular features. A well known example is the peacock: females choose to mate with males with longer, brighter coloured tails, so genes for males with long, colourful tails spread through the population. In the case of orang-utans, it seems that females like to mate with large males, which have big faces, long beards and loud voices, so these traits will have spread, in time, to all male orang-utans.

4

Where are Orang-utans found, and why?

About a million years ago, orang-utans lived throughout much of eastern Asia: from Java in the south, right up into Laos, Vietnam and southern China as far west as the Indian border. Not all orang-utans then were the same as those we see today. They were much more varied, and possibly comprised several different species. Some were much larger than today's orang-utans, and it has been proposed that there was a smaller, lowland form and a larger, montane form—similar to the modern-day lowland and mountain gorillas.

Prime habitats for the orang-utan today include riverine forests in the coastal plains.

23

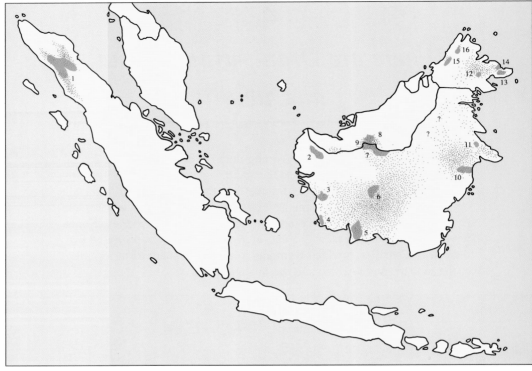

Distribution of orang-utans, and protected area where they occur.

1. Gunung Lauser National Park
2. Gunung Niut Nature Reserve
3. Gunung Palung National Park
4. Muara Kendawangan Nature Reserve
5. Tanjung Puting National Park
6. Bukit Baka/Bukit Raya Nature Reserves
7. Gunung Bentuang dan Karimun Nature Reserve
8. Lanjak-Entimau Wildlife Sanctuary
9. Batang Ai National Park
10. Kutai National Park
11. Sangkulirang Nature Reserve
12. Danum Valley Conservation Area
13. Tablin Wildlife Reserve
14. Kulamba Wildlife Reserve
15. Crocker Range National Park
16. Kinabalu Park

Since then, the range of the orang-utan has shrunk dramatically, and the animals are now only found on the islands of Sumatra and Borneo. Even here, their range is very limited. In Sumatra, orang-utans only occur in the far north, in the province of Aceh. There are occasional, unconfirmed reports from scattered localities elsewhere. The rumours of a legendary, gnome-like figure known as orang pendek or short person in central Sumatra, thought by some people to be an unknown new ape, might be a small, isolated population of orang-utans. Even in very recent times though, the range of the orang-utan in Sumatra has declined: between about 1935 and 1980, it shrunk by 20 to 30%.

In Borneo, the animals are found throughout more of the island, but there are still major gaps in their range. For example, they do not occur in the whole south-east quarter of the island between the Barito and Mahakam rivers, or in the highlands of the Kayan-Mentarang region in Kalimantan,

Tham Nyip Shen

Lowland mixed dipterocarp forests are another important habitat for orang-utans.

in Sarawak north or east of the Rajang River, and they have never been recorded in Brunei Darussalam. In prehistoric times, they did occur more widely in Borneo: large numbers of fossil remains have been found in Niah Caves in Sarawak, way outside the current range of the species. And the range is still shrinking. A hundred years ago, orang-utans occurred throughout much of southern Sarawak, from Kuching in the west to Ulu Balleh in the east. Even 30 years ago, there were records of them occurring in the upper Balleh and upper Baram, but now the animals in Sarawak are restricted to just two areas, and are on the verge of extinction in one of those. Similar range contractions are still occurring in other parts of Borneo.

Plotting a map of exactly where orang-utans occur is almost impossible. This is partly because there is no clear dividing line between forests which have lots of orang-utans, and those which do not have any. Instead, there are large areas with very low numbers of apparently non-breeding animals. On top of that, adolescent and adult males occasionally roam way outside the normal range of the species. Like many humans, such males possibly go through a phase of wandering very widely before they finally settle down and breed.

Within their range, though, orang-utans do not seem too fussy about where they live. They occur in a wide range of habitats, from mangrove forests, freshwater and peat swamp forests, forests in coastal areas near large rivers, right through to steep hills in the interior of Borneo and Sumatra. Though there are anomalies even within that, since there are many peat swamps, mangroves and riverine forests where they are not found. They are also not mountaineers; they do not occur in forests above about 1400 meters altitude, and are found at their highest densities in the lowlands.

So what can possibly have caused such dramatic declines in the range of orang-utans, both in pre-historic and also more recent times? Why is the orang-utan today restricted to such tiny areas, when it used to occur throughout most of south-east Asia? It is not due to climate: many areas

(opposite) Orang-utans breed extremely slowly, which means that even low levels of hunting can cause extinctions. Past hunting has caused the range of the orang-utan to shrink dramatically.

Jimmy Omar

Ku Min Yong

Teeth of past orang-utans are a clue to the present. The scarcity of milk teeth from young orang-utans in fossil remains shows that prehistoric hunters used to kill the adults, and keep the young animals as pets.

where the species no longer occurs still have ideal climates for orang-utans and their forests. It is not just due to habitat loss: large areas of good forest still occur where the animal used to live but does not any more. Within Borneo, one theory is that lack of suitable minerals explains some of the gaps in their range. It does not explain all of them though: there are some areas where orang-utans occur on poor soils with few favourite fruit trees such as figs and durians, and there are forests on good soils where the animals do not live. On top of that, lack of minerals does not explain why orang-utans no longer occur in the large areas where they once thrived.

28

By far the most likely explanation for such wide-scale disappearance of the species, and the one on which most field scientists agree, is hunting. Fossil remains at Niah Caves in Sarawak show that orang-utans there were hunted heavily as long as 35,000 years ago. The earliest human remains in Niah have been dated at 40,000 years old, so almost as long as humans have been in the area, they have hunted orang-utans. Signs of fires, charred teeth and bones, and the presence of large numbers of remains inside the caves implies that people carried the hunted animals to the caves to be eaten. The scarcity of orang-utan milk teeth in the fossils indicates that prehistoric man did much the same as does modern man: he hunted the adults to eat, and kept the infants alive as pets. Amongst the Niah fossils, orang-utans were second in abundance only to pigs, showing that they were a highly favoured food.

Further evidence of hunting is that, as time went on, the size of the orang-utans whose fossils were found at Niah became slightly smaller. This somewhat worrying trend strongly supports the idea that early hunters were killing the larger animals which had more meat. Bigger animals were also likely to have spent more time on the ground, so would have been easier to hunt. This fits well with the idea that the large, prehistoric orang-utans in southern China were also killed by hunting. Even today, although some groups of people in Borneo have taboos against hunting orang-utans, others do not, and yet others actively seek them out for their head trophies. It was thought that this might be a new fad after headhunting was banned, but now people are starting to think that it might be a much older tradition.

If this all sounds somewhat unlikely, bear in mind that hunting rates do not have to be very great to wipe out an animal such as an orang-utan. They are large, slow-moving animals so are easy to kill with a bow and arrow or blowpipe. Their size makes them attractive because one animal provides lots of meat. They occur at low densities, and their breeding rate is one of the lowest known for any mammal. This means that it will take a long time for orang-utans to breed and replace animals killed. Young orang-utans depend on their mothers for at least five years, so killing an adult female will almost inevitably result in the further deaths of one or two offspring.

For any animal species in the wild, you can take information on the number of infants produced every year, and on the number of animals living in the forest, and you can work out the number of animals which can be hunted without driving the species to extinction. For orang-utans in the wild, you can only hunt about one animal per 20 km² per year if the hunting is going to be sustainable. Put another way, if more than one orang-utan is hunted per km² of forest every 20 years, then the population in that area is doomed in the long term. So for such a slow-breeding animal living at such low densities, even extremely low rates of hunting can be devastating.

Hunting, then, does indeed seem to be the main reason why the orang-utan only occurs in such limited areas today. We can even take this further. Some of the top orang-utan field biologists believe that the smaller size, arboreal habits and solitary nature of orang-utans today are all a result of hunting over many thousands of years.

Within the areas where they do still live, the number of orang-utans varies between different habitats. In Borneo, orang-utans reach their highest densities in freshwater swamp forest, where there are about three animals per km² of forest. In lowland forests, there is usually only about one animal per km², and up in the hill forests, the numbers are usually as low as only one animal per three km². This means that there are ten times more orang-utans per unit area in freshwater swamp forests than in the steep hills. In Sumatra, again swamp forests seem to be favoured habitats, and between seven and ten orang-utans live in every km² in the swampy forests of Suaq Balimbing in Gunung Leuser National Park, the highest known density of wild orang-utans in the world. In drier parts of the park, the animals still seem to be more concentrated than in Borneo, up to five orang-utans per km². The reasons for such variations are not clearly known, but are due in part to differences in the abundance of favourite food trees. Gunung Leuser abounds with wild fig trees, one of the orang-utan's all-time favourite foods, so the bounty of fruits allows more orang-utans to survive in the forest.

5
Life History of an Orang-utan

High in the trees, in the dim, cool, pre-dawn light, a female orang-utan gently supports her new infant, born just an hour ago. Eyes tightly shut, it is already clinging hard to its mother's long red fur, and taking its first drink of milk. The infant weighs about 1.5 kg, and has been born after a gestation of about 245 days (just over eight months).

If it survives, the infant can look forward to a long life ahead of it. In captivity, the record for orang-utan life span was held by an orang-utan in San Diego Zoo, which lived for about 59 years. At Sepilok Orang-utan Rehabilitation Centre in Sabah, semi-wild animals are still healthy and fit by age 30, which implies that they live for at least 35 years.

Orang-utans depend on their mothers for many of those years. Our new-born infant will cling to its mother's front continuously for almost the

Young orang-utans depend on their mothers for up to eight years.

Elke Lies

31

P.J. Cribb

Orang-utans almost always give birth to single infants, although very occasionally twins are born. In the wild, the chances of both surviving are small.

W.M. Poon

Young orang-utans spend much of their time in play.

whole of its first year of life, and it will still be riding round on her for much of the time until it is about 2 ½ years old. The youngster will be weaned when it is about 3 ½ years old. By the time it is about five years old, its mother will probably have given birth again, to a new infant. The interval between births for female orang-utans is highly variable. It can be as little as three years, or as long as eight years, but the average is about five to six years. This is amazingly infrequent; it is the slowest breeding rate of any primate, and amongst the slowest of any mammal.

33

Once the young orang-utan has a new sibling, it does not stay as close .to its mother as it did before, and gradually becomes more independent, although it will still turn to her for protection until it is seven or eight years old.

Female orang-utans become sexually mature at about this age, but males do not become fully socially and sexually mature until they are 13 to 15 years old or more. Even then, males will not develop their full adult male attributes of face flanges and throat sacs if they are in the presence of a dominant adult male, so they might have to wait even longer until they are a fully mature, impressive adult male.

Juvenile orang-utans have to watch out for adults who might chase them away from favourite food sites.

S. Von Pettz

6

Social Life of Orang-utans

To say that orang-utans have much of a social life is rather misleading. In the wild, orang-utans are the least social, and least sociable, of any diurnal primate anywhere in the world. For both Sumatran and Bornean orang-utans, adult males almost always live alone, and adult females travel only with their latest infant and possibly one juvenile offspring. Even adolescents usually live on their own, although they occasionally form temporary groups, either just with other contemporaries, or following an adult of either sex for a short time. Otherwise, the only time different orang-utans generally meet is when several animals all pile in to feed at a particularly large fruiting tree such as a huge strangling fig laden with fruits. Even then, the different individuals usually arrive and leave with almost no visible interaction with any of the others.

On the other hand, females at least encounter other animals relatively often. Each female orang-utan lives in a relatively discrete area of forest. The exact size varies somewhat between sites, but is between 60 ha and 500 ha. The ranges of different females overlap, sometimes quite a lot. When a young female grows up and leaves her mother, she is likely to set up her own home very nearby, so the chances are that females in nearby and overlapping ranges are related. So as an adult female roams round her home patch of forest, it will not be very long before she encounters another female, who will either be a relative or another familiar animal. When two females do meet, though, they do not exactly greet each other with open arms, but are more likely to pass by with little or no direct interaction.

Even if related, nearby females are highly unusual amongst primates in that they do not actively seek each others' company, do not groom each other, and do not support each other in interactions with other orang-utans. And they bring up their offspring with no help from others. So the only real bond in orang-utan society is that between a female and her

35

Sub-adult and adult male orang-utans are solitary almost all of the time, and when they meet other males, they confront them aggressively.

own young offspring until they are old enough to move away and fend for themselves. Even so, it is likely that a female does keep track of who is doing what in her neck of the woods, through occasional meetings, calls, and seeing fresh nests.

Adult males are even more antisocial. They actively avoid other adult males, and if ever two adult males do meet, there is almost always an aggressive confrontation. Although fights have only rarely been seen, the large numbers of scars and broken figures on adult males indicate that fighting occurs, and can be fairly violent.

Adult males produce what are known as long calls . These are only made by mature males with full face flanges and throat sacs, and the calls are so loud, they can be heard up to a kilometre away. The call starts as a series of quiet, bubbling grunts, then builds up into a full-blown gravelly roar. It is often accompanied by vigorous branch shaking, and the male looks even larger and more impressive than usual because his hair stands on end during the displays. The long calls seem to have various purposes, and not all are well understood. There is no doubt that one of their main functions is to space out adult males, and to prevent them from bumping into each other inadvertently. A male will always call in the direction in which he is about to travel, and his face flanges possibly help to focus the call that way. He is also most likely to call when he is about to travel a long way, so the call is probably telling the other orang-utans in the new area that he is coming. This is to allow the other males to keep out of his way, especially any younger males who might be around. It might also attract females, and be telling them that a large and handsome mate is in the neighbourhood.

Unlike females, not all males have a discrete patch of forest in which they stay for most or all of their adult lives. Some males do remain in one area for several years or more, and these are invariably fully mature. Such animals are known as resident males, and their ranges overlap those of several females. But then, suddenly one day, a resident male will disappear, leaving his normal home area completely, sometimes for several years before he reappears. The reason for this is unknown but might be because all of the females in his own area have recently produced offspring, so it will be many years before they become fertile again. In which case, his best option might be to develop a wanderlust

37

and go to look for other females further afield. In addition to these resident males, there are also so-called wandering males. These are usually adolescents, but can also be adults. They seem to have no fixed abode, but range over huge areas of forest, often being chased away by resident males if they come too close.

Sumatran orang-utans do tend to be just slightly more social than Borneans, and pairs, or even small aggregations of animals are occasionally seen. This is possibly because, unlike in Borneo, there are tigers in Sumatra which would make a good meal of an orang-utan. So sticking together just that bit more means the animals are more likely to detect a hungry tiger on the prowl, and less likely to become its dinner.

This leads to the question, though, of why be solitary in the first place? The most likely answer is food. Orang-utans are large animals, and need to eat a huge amount. Their main foods are fruits, which are often sparse in the forest, and any one tree is often small, or does not have that much food. So if you have to eat a vast quantity yourself, there is just not enough in any one food source to be able to share it. Which means your only option is to travel alone. The only time you can afford to share is when there is a bumper crop of fruit, and that is when several orang-utans do all feed from the one tree.

The one major exception to this solitary lifestyle is when males and females mate. Sometimes the female mates willingly, but at other times she is obviously an extremely unwilling participant.

Willing matings usually take place during a period of what is known rather charmingly as consortship, and the male is almost invariably a fully mature adult. These consortships are the only times when adults spend significant lengths of time together. The signals to start a consortship can be given by either the adult male or adult female, and the pair then travel around closely together from three to about eight days, although in Sumatra, some consortships will run into months. During that time, the two animals step away from their usual antisocial

---

(opposite) The only close social bond in orang-utan society is that between an adult female and her young offspring.

39

Orang-utans are solitary because food in the forest is so scarce, there is not enough to share with others.

behaviour, and are extremely tolerant of each other, with the male allowing the female to take food from his hand or even his mouth. Matings occur several times during the consortship. Both animals join in willingly, and mating can be initiated by either partner. Even such apparent harmony between two individuals does not mean that they only enjoy one partner. Both males and females mate with different animals at different times, and a receptive female can indulge in consortships with more than one male even during one reproductive cycle. Nonetheless, females are very picky about their choice of partner. They actively solicit certain males, and strongly refuse the advances of others. The most favoured males are usually the older, mature ones with their full magnificence of face flanges, throat sac and beard.

This system would seem to leave a lot of younger and less attractive males out in the cold without a mate. Some adolescent males do form consortships with adolescent females. Many are not satisfied with this though, and try to mate with adult females. Adult females rarely accept this, and are highly unwilling to mate with young males. Undeterred, a young male will often follow a female, and even form a short consortship with her. If the female continues to refuse his advances, his superior size and strength are used to overpower and rape her. Such relationships are either terminated by the arrival of an adult male who displaces the youngster, or else by the female finally managing to get away. Interestingly, there is some evidence that such rapes rarely result in pregnancy, possibly because at the peak period of fertility, the female is most likely to be found with her preferred fully adult male.

Hans P. Hazebroek

7
What do Orang-utan eat?

Being so closely related to us, and of roughly similar size, we might assume that orang-utans like to eat much the same things that we do. Some of their foods we might indeed find rather tasty, especially the sweet succulent fruits, but other parts of their diet would probably make us downright ill.

Orang-utans are predominantly fruit-eaters; indeed, they are the largest animals in the world whose diet mainly comprises fruits. Wild figs are a major favourite, and in forests such as Ketambe in North Sumatra where fig trees are abundant, about a third of the total diet is figs. In other forests where figs are rarer, orang-utans have a more diverse fruit diet. In all areas, though, a single orang-utan will feed on more than a hundred different types of fruit in a year. Many of the fruits are sweet and succulent, such as rambutans and mangoes, so provide lots of energy. Others are more rich in oils and protein such as durians or wild acorns. The orang-utans' strong but extremely agile hands, feet, lips and teeth mean that they can get into any fruit in the forest, however tough its skin, and they also often pull branches and twigs towards them, or even break them off completely, in order to get at that extra bit of fruit just out of reach.

They do eat a lot of other foods as well, especially at times of year when fruits are scarce. So they eat leaves, twigs, bark, honey, and some animal foods such as termites, ants, bees, birds' eggs and small lizards. There are even occasional records of them eating larger animals such as a slow loris, or even a gibbon. Again, in acquiring some of these foods, their strength, manual dexterity and intelligence come into play. Orang-utans are the only animals in the forest which can pull part of the tough, thick bark from a dipterocarp tree, with a combination of hands, feet and teeth,

---

(opposite) To reach tasty fruits, the orang-utan's agility in the forest canopy is critical, as this female with her clinging infant demonstrate.

43

Orang-utans feed mainly on fruits. Particular favourites are: wild figs, mangoes, rambutans (bottom left), Artocarpus (bottom right), and their strong hands even allow them access to the flesh of a tasty durian (top).

William W.W. Wong

William W.W. Wong

Tan Swee Heng

Orang-utans feed mainly in the canopy of trees where most of the succulent fruits are found, but they occasionally come down to gather an extra morsel close to, or even on, the ground.

45

and then use their strong mouths and front teeth to strip off the soft bark lining to provide a tough but varied addition to the diet.

Orang-utans need to eat vast quantities of fruits, so consume hundreds every day, from a selection of different trees and species. On top of that, some of their other foods such as bark lining, honey and insects take quite some time to obtain. This means that almost half of an orang-utan's day is spent just in extracting food and eating it.

Hans P. Hazebroek

This male orang-utan at Batang Ai, Sarawak, has come to a salt lick for the minerals it contains.

46

8

Travel through the Forest

For an animal as large as an orang-utan, obtaining food on small branches at the extremities of rain forest trees, possibly thirty metres or more above the forest floor, presents somewhat of a problem. Yet orang-utans are perfectly adapted to the task: their long arms to stretch out to cling on to the next branch or reach for that distant fruit; their hand-like feet to give them four limbs for holding on and for grasping objects; their strength to pull branches towards them, and their intelligence to work out how best to solve the problem of negotiating the extreme complexities of branches, twigs and lianas to get between A and B, to reach the next crop of succulent figs.

In spite of such adaptations, orang-utans do travel extremely slowly and deliberately through the trees. They never rush, never jump or leap, and all of their movements are careful and cautious. They hold on with at least two limbs almost all of the time, and three or four if possible, to spread their load between different supports, using their feet as well as their hands to grasp branches and lianas. Indeed, their cumbersome way of getting round the canopy has aptly been described as quadrumanual clambering!

To make its way across a small gap in the forest, an orang-utan will hold tightly with all four hands to a flexible sapling, tree or vine, and rock back and forth to swing further and further, until eventually it is close enough to reach out and grab the next tree with its outstretched hand.

Larger gaps are more of a problem for an animal that cannot leap, so both male and female orang-utans occasionally climb down and move between trees along the ground. Again, this is slow and deliberate, largely because an orang-utan's hands and feet are more adapted to life in the trees than on the ground. They walk with all four limbs on the ground, and with their hands made into a clenched fist, unlike the knuckle walking of the other great apes. In captivity, orang-utans

Travel through the forest canopy to reach the next tree or liana of succulent fruits requires extreme agility and skill, and a range of unorthodox climbing techniques.

Albert Teo

C.L. Chan

Albert Teo

C.L. Chan

Travel through the forest is always slow and deliberate, the orang-utan holding on with as many of its four hands as possible at all times.

T.D.Z. Adli

Occasionally, orang-utans will climb down to the ground…

T.D
Ad

And walk along it. In primary forest, this is only common in large adult males which are too heavy to move easily in the tree tops, although in disturbed forests all animals occasionally are forced to the ground to cross large gaps.

C.L. Chan

sometimes walk upright on two legs, but have never been seen to do this in the wild.

Adult male orang-utans, being twice as heavy as females, find travel through the trees even more difficult and hazardous, so come down to the ground somewhat more often. Females travel mainly through the continuous, middle canopy, but males spend more of their time in the lower levels of the forest and on the ground itself. The only orang-utans

C.L. Chan

Very occasionally, orang-utans smother themselves in mud while travelling. The reason for this is unknown.

51

Elke Lies

(opposite and above) Orang-utans build nests to sleep in at night, and occasionally for a nap during the day as well.

which seem more spry in the trees are the young ones, which move about more quickly, and even play by swinging from one arm or leg and occasionally dropping down to the branch below. Such play is not without its risks, and the animals do occasionally slip and fall, though usually scramble quickly back up again.

Orang-utans have to balance their need for slow caution in the trees with their need to reach enough food trees in a day to have enough to eat. The distance that they travel depends on season and availability of food, but is usually between about 300 m and 800 m per day.

In general, it is not easy to tell where in the forest orang-utans have been because, unlike ground-living animals, an orang-utan swinging through the trees leaves no footprints. The one major clue that orang-utans are around, though, is to look for their nests. Because of their size, sleeping high in the trees is somewhat hazardous, so orang-utans build leafy platforms or nests. They usually build a new nest every night, and sometimes during the day too if they want a prolonged nap. As night approaches, the orang-utan selects a suitable sleeping spot—usually a forked branch in the middle of the canopy, with lots of smallish, soft leaves around. The animal then sits in the fork, and folds the nearby branches in underneath itself, breaking off twigs and branches too, to weave roughly into the nest. This makes a concave platform of leaves and branches about a metre across, in which the animal then curls up for the night. Orang-utans are usually very efficient at making nests, and complete them in just a few minutes, although if they take special care, they can sometimes take up to 20 minutes to make a really comfortable nest. Day nests are usually somewhat more flimsy, and flung together more rapidly. True to their solitary nature, orang-utans never share their nests, the only exceptions being mothers with their own young offspring. By about one to two years old, young orang-utans start building their own simple day nests, and by three to four years of age, are making their own night nests.

9
Tools and Language

For many years, orang-utans were thought to be the least intelligent of the great apes: slow, lumbering, solitary animals, that did not possess the tool-using or language abilities of the much quicker chimpanzees and gorillas. Recent research has now shown that this image is wrong. Orang-utans are extremely intelligent, and their tool using and language skills are easily on a par with those of the African great apes.

It has long been known that orang-utans made simple tools, but so basic, they were not considered remarkable. In a heavy rain storm, they will pluck off a large leaf to use it as an umbrella. If they have an itch in the middle of their backs, they will use a stick to give it a good scratch. And anyone who has watched orang-utans in the forest knows that when an

Recent research shows that, far from being the least intelligent of the great apes, orang-utans are extremely clever, with well-developed tool-making and language skills.

Orang-utans are imaginative in selecting and using twigs as tools, for example, to extract honey or insects from holes in trees.

orang-utan gets agitated at the human observers below, he will hurl a branch down at them with unnerving accuracy.

Recent research by Carel van Schaik in Sumatra, though, has shown that this is just the tip of the iceberg in terms of orang-utan tool using ability. In the deep swamp forests of Suaq Balimbing in the south of Gunung Leuser National Park, orang-utans have developed a range of tools that rival those of chimps: 54 different tools for extracting insects, and 20 for opening or preparing fruits. For example, to obtain honey from a hole in a tree, an orang-utan will break off a small branch, strip the bark from it, and poke it into the hole, extracting it with a glob of gooey honey on the end. Without the tool, the honey would have been out of reach. Similarly, orang-utans use tools to extract termites from their nests, or to remove the stinging hairs from fruits so that they can be eaten.

Orang-utans are also now known to be able to learn simple language, and exhibit traits regarded as intelligent such as simple sentence construction—and even lying!

Ku Min Yong

Orang-utans use simple tools such as this adult female cleaning her teeth with a bent leaf stem.

Each tool is custom made for its task. A honey-extracting stick is small, and with the bark peeled off. A termite-extracting stick is larger and with its bark intact. If a tool is too long, too short or too fat, it is thrown way. If it is too long, the orang-utan shortens it.

If orang-utans have such tool making abilities, why has nobody seen them before, in other sites where orang-utans have been studied for many years? One possibility is that because Suaq Balimbing is swamp forest, insects nest in holes in trees, not on the ground, so they are more

58

inaccessible. Another possibility is that since the population density of orang-utans at Suaq Balimbing is higher than at any other site, animals must spend more time relatively near to others. This means that there are more opportunities for orang-utans to learn, and to pass on their tool making skills: when an orang-utan is using a tool, if there is any other orang-utan nearby, it inevitably comes over to watch. In any case, the Suaq Balimbing animals have proved that orang-utans have an ability and problem-solving intelligence far beyond what was originally supposed.

The two attributes which used to be thought uniquely human were our ability to make tools, and to use language. All of the great apes are now known to be tool makers and users. For the past twenty years, we have also known that both chimpanzees and gorillas can learn language. They cannot use spoken language, but that is because they do not have the right shaped throats, mouths and voice boxes; it is not due to their lack of intelligence. In captivity, they are capable of learning American sign language, to a fairly sophisticated level. But what about orang-utans? They are much less social than the other great apes, so are their communication skills also lower?

Work with one captive young male orang-utan at the University of Tennessee has shown the answer to be no. The orang-utan was taught American sign language, and eventually learned to use about 150 symbols. His language abilities were similar to a human child of about two to three years old. He could use signs for people, names, places, foods, actions, objects, pronouns and attributes. He could make his own simple constructions, such as signing "bad bird" at noisy birds giving alarm calls, or "eye drink" for the contact lens solution used by his keeper. He could understand simple vocal speech, and reply in his own signs. He could tell lies; somewhat ironically, this is considered a sign of intelligence since it assumes that the liar can visualize another individual's viewpoint and reactions. When playing, for example, the orang-utan took a pencil eraser, pretended to swallow it, opened his mouth to show it was not there, while signing "food-eat". In fact, he had hidden it in his cheek, and later took it out.

He could also ask for objects which were not present at the time. Again, this is significant; conceptualizing things not there is considered an

The orang-utans' intelligence is not surprising, given their ability to remember and even predict the location of food trees in a complex, diverse rain forest.

S. Von Peltz

important attribute in the development of language, since it allows discussion of distant objects and events. In the case of an orang-utan, this ability should have been no surprise, given a wild orang-utan's complex knowledge of the forest, and ability to remember and even predict which distant trees will be coming into fruit and when. Indeed, some people think that the need to be able to predict the constantly changing patterns of food distributions in a highly complex rain forest is one reason for the orang-utan having developed such intelligence.

Whatever the reason for its evolution, these recent studies have confirmed that orang-utans are indeed highly intelligent. It used to be thought that the intelligence of humans, chimpanzees and gorillas evolved partly because of the pressures of complex social lives and relationships. The abilities of orang-utans, with their much more solitary lives, cast doubt on this, and we must look in future for other explanations.

10
Conservation of Orang-utans

THREATS FACING ORANG-UTANS

The range of the orang-utan has shrunk dramatically, from once covering much of eastern Asia, to now being restricted to limited parts of Borneo and Sumatra. Historically, the single biggest reason for such declines was probably hunting, but does hunting of orang-utans still continue, and is it still the greatest threat facing the species? Indeed, is the orang-utan still threatened, given our awareness of its needs and concern for its future?

Under the law, the orang-utan is totally protected throughout its range. This means that it is illegal to hunt orang-utans, to sell them, or keep

Under both national and international wildlife laws, the orang-utan is totally protected throughout its range.

E.L. Bennett

In the past, hunting using traditional weapons such as this Penan blowpipe was the main cause of declines and local extinctions of orang-utans.

E.L. Bennett

Now, most hunters use shotguns. Hunting of primates is common throughout Borneo, as shown by these pig-tailed macaques shot by Ibans in Sarawak. Orang-utans are only hunted rarely, but are still shot for their skulls as trophies, and mothers are killed to obtain their infants for the pet trade.

them in captivity. In addition, there is an international trade regulation, known as CITES, or the Convention on International Trade in Endangered Species of Wild Flora and Fauna. The orang-utan is listed on Appendix I of CITES, which gives it the highest level of protection, and means that all international trade is banned, except for non-commercial scientific purposes such as captive breeding.

In theory, then, the orang-utan ought to be well protected, but sadly this is not the case. Orang-utans are occasionally hunted to provide skulls for sale to tourists, some of which are elaborately carved. Others are deemed to have been hunted by traditional headhunters, which is unlikely to be true, but if so is still a drain on the wild population. Orang-utans are also hunted by some local people, for food, for sport, and as crop pests. None of this hunting is large-scale, however, and at present does not form a major threat to the species, except in certain local areas.

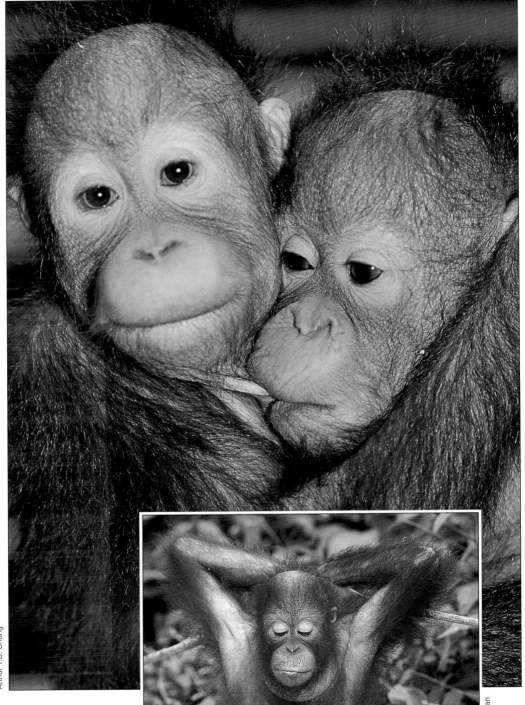

Arthur Y.C. Chung

C.L. Chan

C.L. Chan

C.L. Chan

(opposite and above) Young orang-utans are extremely playful and highly appealing, so the demand for them as pets will always be a problem.

Far more of a problem in recent years has been taking animals from the wild as pets. Young orang-utans are incredibly appealing, and television shows and films featuring orang-utans have led to an international demand for them. The effect of this is worse than might initially appear. No orang-utan mother is going to let her infant be taken away, so the only way to obtain the youngster is to shoot the mother. The conditions under which the animals are then smuggled out are usually so bad that for every young animal which arrives at its destination, several others will have died. It has been estimated that each live pet means six to ten dead animals along the way. Trade is generally small scale but occasionally turns into a bigger problem. For example, a sudden and major trade into Taiwan sprang up in the late 1980s. It followed a television programme featuring a young orang-utan, and the subsequent demand for pets was so great, the authorities could not react in time. Up

Sylvia Yorath

The single greatest threat to orang-utans today comes from land clearance—for farms such as this.

to a thousand orang-utans were smuggled into Taiwan before they could clamp down and stop the trade. This probably meant a loss of at least six thousand animals from the wild. In Taiwan, the trade was successfully halted by changing the law, strict enforcement and public education, but it shows what can happen if controls slip for even a short time.

With some exceptions, hunting of orang-utans today is largely under control. But strict and constant vigilance is needed by law enforcement agencies in Malaysia and Indonesia where the animals occur in the wild, and in countries where pets and skulls might be heading. Once this relaxes, the appeal of the animals will inevitably mean that a trade will once again start, and its effects can be rapid and highly damaging. Nonetheless, at present hunting is generally not the threat to orang-utans that it once was.

On the other hand, orang-utans today face a totally different threat, and that is, loss of their forest home. In 1960, most of Borneo and Sumatra were still covered by rain forest. Now, Malaysia and Indonesia have two of the fastest growing economies in the world, and their human populations are expanding rapidly. This inevitably results in a need for land. Oil palm, rubber and cocoa plantations, housing areas, factories and transmigration schemes all involve clearing huge areas of forest. To take Sabah as one example, in the late 1970s, 30% of the total area of the State was deemed to be suitable for commercial agriculture. Since then, much of this has already been felled and planted, and plans are underway to clear the rest. Virtually all of this is in the eastern lowland and swamp forests: prime habitats of the orang-utan. This picture is being repeated across the animal's range.

Large areas of forest do remain, but much of it is allocated for timber production so is being selectively logged. So can the orang-utan survive in logged forest? No developing country wants to tie up huge areas of land in reserves which produce few economic returns, so if forests can produce revenue from timber, and also be a home for orang-utans and other wildlife, it would be a good compromise all round. Orang-utans can certainly survive in logged forests, but it is not clear if they breed well there. So they might not still be there in the long term. Orang-utans live for more than 30 years, so would continue to be seen long after a forest has been logged, even if they are not breeding. Results from

E.L. Bennett

Orang-utans might be able to survive in selectively logged forest, provided that felling intensity is light, hunting does not occur and, very importantly, large areas of primary forest "islands" are left uncut.

68

different field studies contradict each other, probably because logging practices and damage levels vary greatly between areas. It is known that orang-utans move away from the immediate disturbance, and often end up in small pockets of primary forest, which become over-crowded. It is also clear that intact forest patches are needed to allow populations to survive in the long term. So if large primary forest islands can be left within logging concessions, then logged forests could be a valuable habitat for orang-utans. And since such timber-producing forests will inevitably cover much larger areas than strict reserves, it will allow more orang-utans to survive than would otherwise be the case.

Ultimately, though, the only places where orang-utans will be truly safe is in totally protected areas such as national parks, wildlife sanctuaries and nature reserves: they are the only areas where conservation is the top priority. Fortunately, the number and size of reserves where orang-utans occur has increased greatly in recent years. Between Sumatra, Kalimantan, Sabah and Sarawak, there are now about 15 reserves with significant populations of orang-utans. This might sound a lot, but there are still problems. In some reserves, large areas are unsuitable habitat for the animals: approximately two-thirds of Gunung Leuser National Park is at too high an altitude to contain many orang-utans, for example. Also, few if any current reserves are totally secure against encroachment and hunting.

So what can we conclude? Hunting of orang-utans continues, for food, sport, as crop pests, and for the trophy and pet trade. Public awareness and law enforcement mean that it is not a major problem in most areas at most times, but constant vigilance is needed to prevent it getting out of hand. By far the biggest single problem facing orang-utans today is loss of habitat, mainly from large-scale forest felling for agriculture. The current system of reserves is good and improving, but many prime orang-utan areas are still outside the system and destined for clearance. Some existing reserves are too small to protect orang-utan populations in the long term, and many are subject to at least some degree of encroachment and poaching. Improved law enforcement, increasing the area of land under reserves, effective protection of those reserves, and increasing the value of logging areas for wildlife are all urgently needed if the forest-dwelling red ape of Southeast Asia is to be guaranteed a safe passage into the second half of the twenty-first century.

How many orang-utans are left in the wild?

Nobody knows how many orang-utans there are left in the wild—surprisingly, perhaps, for a large animal in which so many people are interested. This is because it is impossible to survey orang-utans without spending vast quantities of time, manpower and money. Bear in mind that they live high in the trees in dense rain forests, at very low densities, so the chances of even seeing one are small. Many of their forests are either in deep swamps or up steep hills and mountains, so surveying them is difficult and slow. Some forests have good populations of orang-utans, but other forests which seem similar either have very low numbers or none at all. So you cannot take figures from one area and assume that they apply elsewhere. The only way to know if there are orang-utans in a forest is to go and look, yet Borneo and Sumatra contain some of the most remote and inaccessible regions in the world.

People always want to know numbers though, and over the years, estimates have been made. Sumatra is somewhat easier than Borneo, because the animals are confined to a more limited area. Even so, it is impossible to be precise. The best estimate was made in 1979, when orang-utan expert Herman Rijksen said that there were between 5,000 and 15,000 animals left. Given the 18 years which have passed, and a probable 20% decline in numbers due to forest loss, there are now probably between 4,000 and 12,000 orang-utans remaining in Sumatra.

For Borneo, opinions have been far more variable, given that we simply do not know if the animals occur in some parts of the island, and the extensive areas where just the occasional one is seen. So estimates for Bornean orang-utans in recent years have ranged from 10,000 to 100,000 animals! Given the very rough known areas of habitat, there are likely to be about six times more orang-utans in Borneo than Sumatra, which would give us a figure of somewhere between 24,000 and 72,000 animals for Borneo. This fits with possible figures for Sabah and Sarawak, but with no figures available at all for Kalimantan, these can only be guesstimates at best.

Nobody knows how many orang-utans there are left in the wild today.

Rather than absolute numbers, it is possible to assess *relative* numbers of orang-utans in an area by counting their nests. If the number of nests in one area is ten times higher than in another area, then we know that there are roughly ten times more animals there. Most important, if the number of nests in an area goes down over time, we know that the orang-utan population is decreasing. So in terms of conservation, what matters most is not the actual number of orang-utans, but the fact that from nest counts, from the amount of habitat every year which is being cleared, and the numbers of animals lost to hunting and the pet trade, we know that populations are still declining rapidly.

ORANG-UTAN REHABILITATION: DOES IT WORK?

If the wildlife authorities are to enforce the law and the wildlife trade is to stop, they must confiscate orang-utans from traders and pet owners. This gives the authorities a problem though: what to do with the confiscated animals? And we are talking about a lot of animals. In Taiwan alone, in 1993 the wildlife authorities were looking after almost 300 orang-utans which had been confiscated from pet owners. On top of that, the rapid clearance of forests is causing numerous orang-utans to lose their homes. Many end up in islolated forest pockets, and responsible plantation mangers call in the wildlife authorities to rescue them before the patch is cleared. Again, numbers are large: in Sabah alone, more than 200 animals were rescued from forests being felled between 1993 and 1996.

It is usually prohibitively expensive for the wildlife authorities to keep the animals themselves. It costs approximately US$ 8,000.00 a year to keep an orang-utan in captivity, if it is to be given good housing, food, maintenance and veterinary care. Wildlife authoritites all over the world are always short of money, so if they have to spend large amounts on

E.L. Bennett

A major problem facing wildlife authorities is what to do with confiscated orang-utans.

Even after they have been returned to the forest, young rehabilitant orang-utans have to be fed—in this case, with milk.

Rehabilitation centres such as Sepilok, Sabah, are widely seen as the most humane solution on what to do with former captive orang-utans. Here, released animals are returning from the forest to a feeding platform to eat.

A controversial question is whether orang-utan rehabilitation centres should be open to the public. On balance, the need for public awareness and support for conservation means that some should be; centres are a major attraction for local people and tourists, and provide excellent opportunities for education.

keeping confiscated animals, then the amount for protection of orang-utans in the wild is correspondingly less.

So what should be done with these animals? The obvious answer is to find an area of forest, and release the orang-utans back into it. Unfortunately, this is not as easy as it appears.

The first hazard is the risk of introducing disease to wild populations. Orang-utans are so closely related to us, they are susceptible to many of the same diseases. Animals which have been in close contact with humans can pick up diseases and parasites which they pass on to wild populations, and can devastate them if they have no resistance. The second problem is introducing more animals than a forest can support. In the wild, the population of any species naturally increases until it is limited by a resource such as available food or breeding sites. The maximum number of animals the area can support is known as the carrying capacity. If additional animals are released in an area, there will not be enough food for both them and the resident animals. Eventually either the introduced animals or some of the residents will die until carrying capacity is again reached. So introducing more animals will not increase the number of animals in the wild, and will lead to a slow death through starvation of either the released or the resident animals.

The obvious alternative is to release the orang-utans into an area where there are no wild ones. Problem solved? No. If a species does not occur in an area, there is a reason why not. The most likely ones are lack of suitable habitat, or hunting. If the former, then released animals wil not survive for long. In the early 1960s, orang-utans were released into Bako National Park in Sarawak. Orang-utans need large areas of tall forest, which do not exist at Bako, and the animals died. If orang-utans no longer occur in an area because of hunting, then the hunting must be stopped before any release is attempted, otherwise released animals will be killed.

Another problem with highly intelligent animals such as orang-utans is that so much of their behaviour is learnt. Orang-utans spend their first five years with their mothers, learning the skills needed to survive in the forest. Animals which have spent most of their lives outside the forest do not know how to survive in it, and a long and costly learning period is needed.

So what can be done? It seems like a Catch-22: animals have to be confiscated to enforce the law, or to save animals doomed by forest clearance. Yet it is too expensive for wildlife authorities to look after them in captivity, and releasing them into the wild can lead to suffering, and damage to remaining wild populations.

Rehabilitation centres for orang-utans are widely seen as the only option, and have been set up in all parts of the species' range. Over the years, as people have become more aware of the problems, the ways in which animals have been rehibilitated have been improving, For example, a new programme in Wanariset, near Balikpapan in Kalimantan is attempting to overcome some of the problems which have beset rehabilitation centres in the past. All animals, and also all humans, are given intensive medical screening and care, to reduce risks of disease transmission. Orang-utans are only released into forests where there are no resident populations, so none are put at risk. All forests are studied before animals are released to ensure that there are enough food trees to support them. Such a programme is slow and expensive, but is the only long-term option which deals with the problem of confiscated animals, while not jeopardizing the future of the few remaining wild populations.

One big question mark remains: should these rehabilitation centres be open to the public? The presence of visitors hampers rehabilitation. If there is any contact between visitors and orang-utans, it increases risks of introducing disease to the animals, and greatly slows down their becoming less dependent on humans. On the other hand, rehabilitation centres, if run well, are the best possible way of teaching people about orang-utan conservation: seeing the animals in their natural habitat, and learning about them, their problems and needs. Such education is critical if, in the long term, we are going to care and understand enough about orang-utans to want to protect them. And people caring and understanding are, ultimately, the only hope for any species. So allowing people in to certain centres, keeping them some distance from the animals, but providing good interpretation, displays, talks and other information maximizes the education value of the centres, and the conservation value of the confiscated animals.

With these centres, however, we must never lose sight of the fact that the ultimate goal of conservation is protection of the species *in the wild*.

Rehabilitation programmes, if properly run, are a humane way of dealing with confiscated animals, and which give at least a chance that they might one day contribute to a new wild population. But ultimately, the only way to guarantee survival of the orang-utans in the wild is full protection of whole populations of truly wild animals. Ensuring that orang-utans are never taken out of the forest in the first place, and that those forests are protected, is the only guarantee of the species surviving.

Ultimately, the goal of all rehabilitation centres is for former captive animals to return successfully to the wild, in a way which enhances protection of the species as a whole, not just the welfare of individual animals.

What can I do about it?

People often feel frustrated about wildlife conservation. They want to help, but do not know how. So here are a few simple things which everyone can do, which will help to save the orang-utan in its forests of Borneo and Sumatra:

- Never buy an orang-utan as a pet. Even if you feel extremely sorry for one when you see it in a market, do not buy it. Doing so will only mean that the traders make money, they will know there is a market for the animals, so will replace your animal with another. This means you buying your pet will result in another female being shot to get another infant... and the cycle continues.

- Do not buy orang-utan skulls or other trophies. Whatever the dealer says, the skull is unlikely to be old, and your buying it will encourage further killing of animals for future trade.

- If ever you see or hear of an orang-utan or orang-utan trophies being offered for sale, please report it immediately to the wildlife authorities. They can then confiscate the animal or trophy, and will hopefully prosecute the trader. This is the best way to ensure that the trade does not continue.

- Tell your friends and relations. Teach them about the value of orang-utans and other wildlife, and encourage them never to buy wildlife pets and other trophies. Enjoy wildlife with them; go with them to good rehabilitation centres, national parks, nature reserves or other natural areas to learn about and enjoy wildlife in its proper home: the wild.

C.L. Chan

79

C.L. Chan

Lower Kinabatangan River, Sabah.

11

Where to see Orang-utans in Borneo and Sumatra

Truly wild orang-utans are generally extremely difficult to see. They occur at low densities, their forests are often either in deep swamp or up steep, rugged hills, and many of the forests where they occur are remote, inaccessible, and often not open to the general public.

Seeing orang-utans in the wild can be a thrill which you will remember for the rest of your life though, and there are some places to see them much more easily. Below are some of the most accessible sites, where you stand a good chance of seeing orang-utans.

First, there are the rehabilitation centres. The orang-utans here are former captives, but are now semi-wild, and you see them in their natural forest. This means that you can totally guarantee seeing them, even on a short trip, and the animals are in the forest, but also close enough so that you can take good photographs. The centres which are open to the public are:

Sepilok, Sabah

This is the most accessible of all the centres. It is about 20 km outside Sandakan, just off the road to Kota Kinabalu and Lahad Datu. Sepilok lies at the edge of the 43 km² Sepilok Forest Reserve. It is well worth a visit from many points of view. First, views of orang-utans coming in from the forest to the feeding stations are dramatic, and provide golden opportunities for photography. Second, Sepilok comprises beautiful tall, primary lowland dipterocarp forest which has become so rare elsewhere, and walks through the forest are very rewarding in terms of the spectacular trees, and the chance to glimpse a myriad of rain forest animals. To visit Sepilok, take a taxi or bus from Sandakan to the centre; buses are frequent. You do not need a permit in advance, just pay a small

Sepilok Orang-Utan Rehabilitation Centre in Sabah is the most accessible of all of the orang-utan centres.

entrance fee on arrival. Many tour companies also organize trips there. The Rainforest Interpretation Centre (RIC) nearby is also worth a visit, although you need to arrange this in advance through the Forest Research Centre, or else a local tour company could help you to do so.

Tanjung Puting National Park, Central Kalimantan

This large national park (3,040 km²) contains wild orang-utans, as well as many rehabilitant ones which are easy to see in the forest around Camp Leakey. The park comprises a mixture of freshwater swamp and heath forests, and has a huge variety of other animals. Especially easy to see are the proboscis monkeys, as well as a superb array of bird life such as oriental darters, hornbills and kingfishers. To visit the Park, fly to Pangkalanbun; there are daily flights from Jakarta, Pontianak and Banjermasin. If you are going with a tour company, they will organize all your permits and transport for you. If not, then in Pangkalanbunyou need to obtain a walking permit for the Park from the Police Office. (Foreigners need a photocopy of the photo and visa pages of their passport.) Then take a taxi or colt to Kumai. In Kumai, you obtain your park entrance permit from the PHPA office in Jalan Idris (the shoreline road). Local hotels will often arrange the permit for you. From Kumai to the park is by boat. Cheapest is to take the daily water taxi to Aspai, but many people prefer to charter their own longboat or kelotok. Once at the park, there are various hostels where you can stay, and which will arrange guides and transport for you within the park.

Bohorok, North Sumatra

Bukit Lawang at Bohorok is by far the best place to see Sumatran orang-utans. Bohorok is set in the enormous Gunung Leuser National Park (8,080 km²), which comprises beautiful primary lowland, hill and montane forest, and is one of the most outstanding national parks in Asia. It contains both rehabilitant and wild orang-utans, as well as a vast wealth of other Sumatran rain forest wildlife, such as lar gibbons, siamang, the rare Thomas' langur, and a superb array of hornbills and other frugivorous birds. Forest trails are good. To visit Bohorok, fly to Medan, then take a bus or a taxi for the 2-hour drive right into Bukit Lawang. Permits for the Park are available from the PHPA office in Jalan Juandang Boga in Medan, or PHPA office in Bukit Lawang. There are hostels and bungalows in the village where you can stay, and camping is available in the Park. (The other rehabilitation centre in Gunung Leuser National Park, Ketambe, was stopped in 1979, and the site is no longer open to the public.)

Semonggoh and Matang Wildlife Centre, Sarawak

At the time of writing, the rehabilitation centres for orang-utans in Sarawak are changing. The long-established Semonggoh is likely soon to be closed to the public, since the forest reserve is too small for successful rehabilitation, so the centre will become a veterinary and research centre only. Instead, Matang Wildlife Centre will be opened in mid-1998. This is in Kubah National Park (22 km²). At present, orang-utans will not be rehabilitated here, due to the small size of the park, lack of suitable habitat and hunting. (This might change in future, if the park can be extended and hunting controlled. Plans for both are underway.) Instead, confiscated orang-utans will be kept in a large enclosure of natural forest, with excellent viewing conditions. The centre has a good

C.L. Chan

Semonggoh Rehabilitation Centre is easily accessible from Kuching, Sarawak.

interpretation centre, and comfortable accommodation. Forest trails and waterfalls are other attractions. To reach Matang, take a taxi or bus from Kuching directly to the centre. You can obtain your entrance permit on arrival. If you wish to stay overnight, you need to obtain a permit in advance from the National Parks Booking Office in the Tourist Information Centre, next to the Sarawak Museum in Kuching.

For those who wish to see truly wild orang-utans away from the rehabilitation centres, there are four options where access is relatively easy. The chances of seeing orang-utans varies between them, and also between seasons, so you need patience, time, and to walk extremely quietly in the forest—and you also need good luck!

Borneo Rainforest Lodge, Danum Valley, Sabah

This is probably the best place anywhere for seeing truly wild orang-utans. The Rainforest Lodge is set in the 438 km² Danum Valley Conservation Area, which in turn is part of a much larger block of surrounding forest concession. The forest at Danum is tall, primary,

84

lowland forest, and the absence of hunting means that it is the best place in northern Borneo to see a full array of lowland tropical wildlife. Borneo Rainforest Lodge comprises high quality accommodation, food, and transport, with good guides, so is expensive. If you can afford it though, it is well worth it, since the opportunities for viewing wild orang-utans in primary forest are uniquely good, and if you stay for two to three nights, you stand a very good chance of seeing them. You can also guarantee seeing a wealth of other forest animals, and there is a spectacular canopy walkway. To book, contact Borneo Rainforest Lodge, P.O. Box 11622, 88817 Kota Kinabalu, Sabah. To reach the lodge, fly to Lahad Datu where you will be met and taken in by road, about a two hour ride.

Borneo Rainforest Lodge, Danum Valley, Sabah.

85

Danum Valley Field Centre, Sabah

This is also at Danum Valley, on the opposite side of the Conservation Area to the Lodge. Hence, opportunities for seeing orang-utans and other wildlife are also excellent. The centre is not generally open to the public, but can be visited by researchers with permission, and also school parties and other educational groups. To find out if it is possible for your group to visit, contact Danum Valley Field Centre, P.O. Box 11622, 88817 Kota Kinabalu, Sabah.

C.L. Chan

Danum Valley Field Centre, Lahad Datu, Sabah.

Lower Kinabatangan River, Sabah

This area is a mosaic of riverine forests, set along the Kinabatangan River. It has a wealth of wildlife which is easy to see, including proboscis monkeys, macaques, red langurs, Hose's langurs, and many superb water birds such as oriental darters, kingfishers and egrets. The area also has wild orang-utans. Sightings in the area are relatively frequent, and are greater in certain seasons. (September and October are probably the best.) But, as with wild orang-utans anywhere, a good deal of luck is involved. It is worth visiting anyway though, as it is the best place in Malaysia for seeing proboscis monkeys and such a high density of other primates. Several Sandakan-based tour companies have their own lodges there, so the easiest way to visit is to book through them, and they will arrange transport, including boats to cruise along the rivers looking at wildlife.

Batang Ai National Park, Sarawak

This area contains the highest density of orang-utans in Sarawak, both inside and outside the Park. Nonetheless, the terrain is steep, and chances of seeing the animals vary depending on the time of year. With a good guide, you will certainly see nests, and possibly hear male long calls. Time, energy and good luck need to be on your side to see the animals. The area is well worth visiting though, for its beautiful forests and scenery, and for the Iban longhouses with some of the richest culture in the region. The best way to visit the area is through one of the several local tour operators in Kuching who have their own lodges up there, and who will arrange transport and guides for you. Access is by road from Kuching to the Batang Ai dam (approx. 4 hours), then upriver by longboat (a further 2–3 hours).

In Indonesia, there are also places where you can view wild orang-utans, although they are generally more difficult to reach and you need to be more adventurous. The most accessible sites where you have a good chance of seeing them are the areas of Tanjung Puting and Gunung Leuser National Parks away from the rehabilitation centres, so make your way to Camp Leakey or Bohorok, then consult your local guides. If you want to go further off the beaten track, there are many opportunities, and your best option is to consult one of the good guidebooks for travel in Indonesia (e.g., Periplus or Lonely Planet guides).

Acknowledgements

For this book, I have drawn heavily on the material of field biologists who have studied orang-utans for many years in the wild. In particular, the writings of the following people have been crucial in providing much of the data on which the book is based: Dr John MacKinnon, Dr Biruté Galdikas, Dr Junaidi Payne, Dr Herman Rijksen, Dr Peter Rodman and Dr Carel van Schaik. Dr Margaret Kinnaird and Dr Tim O'Brien took a lot of time to dig out the information on seeing orang-utans in Indonesia, for which I am very grateful.

Special thanks go to Dr Ken Searle and Karen Phillipps, who provided me with a wonderful working environment, generous hospitality and use of their excellent library while writing the book. I would also like to thank the Sarawak Forest Department, especially Datuk Leo Chai, Cheong Ek Choon, Sapuan Ahmad, Melvin Gumal and Oswald Braken Tisen for supporting my work in Sarawak over the years.

Particular thanks go to Chan Chew Lun of Natural History Publications (Borneo) Sdn. Bhd. for inviting me to write the book, and for his unfailingly cheerful good patience in waiting for it to appear. I would also especially like to thank all of the photographers who have allowed their pictures to be used, and which are the key element to making this an attractive book: Tengku Dr D.Z. Adlin, C.L. Chan, Arthur Y.C. Chung, Dr Phillip Cribb, Drs Hans P. Hazebroek, Ku Min Yong, Elke Lies, Therese Lillieskold, Aaron J.L. Law Jimmy Omar, Stephen Von Peltz, Dr Andrew Plumptre, W.M. Poon, Tan Swee Heng, Albert Teo, Y.B. Datuk Tham Nyip Shen, William W.W. Wong and Sylvia Yorath. Others whose help in producing the book is appreciated are Grace Tsang, Chua Kok Hian, Goh Shee Onn, Charles Yeoh, Lee Seok Chu, and, for producing the map, Yong Ket Hun.

My work is fully funded by the Wildlife Conservation Society (formerly New York Zoological Society), and I would particularly like to thank all of the people in WCS who continue to support me, as well as to be valued colleagues and friends, especially Dr John Robinson, Martha Schwartz, Dr Alan Rabinowitz and Dr Joshua Ginsberg.

Suggested Further Reading

Many people have written about orang-utans, and possible further reading ranges from popular accounts by people who have worked in the field with them, through to full academic monographs. Below are just a few pointers for the general reader interested in orang-utans, and the natural history of Southeast Asia.

Bennett, E.L. & F. Gombek, (1993). *Proboscis Monkeys of Borneo*. Natural History Publications (Borneo), Kota Kinabalu.

Campbell, E.J.F. (1994). *A Walk through the Lowland Rain Forest of Sabah*. Natural History Publications (Borneo), Kota Kinabalu.

Cubitt, G. & J. Payne, (1990). *Wild Malaysia*. New Holland, London.

Cubitt, G., T. Whitten, & J. Whitten, (1992). *Wild Indonesia*. New Holland, London.

Francis, C.M. (1984). *Pocket Guide to the Birds of Borneo*. The Sabah Society, Kota Kinabalu and WWF Malaysia, Kuala Lumpur.

Galdikas, B.M.F. (1995). *Reflections of Eden: My Life with the Orangutans of Borneo*. Victor Gollancz, London.

MacKinnon, J. (1974). *In Search of the Red Ape*. Collins, London.

MacKinnon, J. & K. Phillipps, (1993). *A Field Guide to the Birds of Borneo, Sumatra, Java and Bali*. Oxford University Press, Oxford, New York and Tokyo.

Payne, J. & M.Andau, (1989). *Orang-utan: Malaysia's Mascot*. Berita Publishing, Kuala Lumpur.

Payne, J., C.M. Francis, & K. Phillipps, (1985). *A Field Guide to the Mammals of Borneo*. The Sabah Society, Kota Kinabalu and WWF Malaysia, Kuala Lumpur.

Other titles available through *Natural History Publications*

Mount Kinabalu: Borneo's Magic Moutain—an introduction to the natural history of one of the world's great natural monuments
by K.M. Wong & C.L. Chan

Kinabalu: Summit of Borneo
edited by K.M. Wong & A. Phillipps

Kinabalu: The Haunted Mountain of Borneo (Reprint)
by C.M. Enriquez

A Colour Guide to Kinabalu Park
by Susan K. Jacobson

The Larger Fungi of Borneo
by David N. Pegler

Mosses and Liverworts of Mount Kinabalu
by J.P. Frahm, W. Frey, H. Kürschner & M. Menzel

Pitcher-plants of Borneo
by Anthea Phillipps & Anthony Lamb

Nepenthes of Borneo
by Charles Clarke

Rafflesia: Magnificent Flower of Sabah
by Kamarudin Mat Salleh

Tree Flora of Sabah and Sarawak Vol. 1
edited by E. Soepadmo & K.M. Wong

Tree Flora of Sabah and Sarawak Vol. 2
edited by E. Soepadmo, K.M. Wong & L.G. Saw

The Morphology, Anatomy, Biology and Classification of Peninsular Malaysian Bamboos
by K.M. Wong

The Bamboos of Peninsular Malaysia
by K.M. Wong

The Bamboos of Sabah
by Soejatmi Dransfield

Rattans of Sabah
by John Dransfield

Orchids of Borneo Vol. 1
by C.L. Chan, A. Lamb, P.S. Shim & J.J. Wood

Orchids of Borneo Vol. 2
by Jaap J. Vermeulen

Orchids of Borneo Vol. 3
by Jeffrey J. Wood

Slipper Orchids of Borneo
by Phillip Cribb

Birds of Mount Kinabalu, Borneo
by Geoffrey W.H. Davison

Proboscis Monkeys of Borneo
by Elizabeth Bennett & Francis Gombek

A Field Guide to the Mammals of Borneo
by J. Payne, C. Francis & K. Phillipps

The Systematics and Zoogeography of the Amphibia of Borneo
(Reprint)
by Robert F. Inger

The Natural History of Amphibians and Reptiles in Sabah
by Robert F. Inger & Tan Fui Lian

A Field Guide to the Frogs of Borneo
by Robert F. Inger & Robert B. Stuebing

Pocket Guide to the Birds of Borneo
by Charles M. Francis

Birds of Pelong Rocks
by Marina Wong & Hj. Mohammad bin Hj. Ibrahim

The Fresh-water Fishes of North Borneo
by Robert F. Inger & Chin Phui Kong

Marine Food Fishes and Fisheries of Sabah
by Chin Phui Kong

Termites of Sabah
by R.S. Thapa

Common Seashore Life of Brunei
by Marina Wong & Aziah binte Hj. Ahmad

Common Lowland Rainforest Ants of Sabah
by Arthur Chung

Borneo: the Stealer of Hearts (Reprint)
by Oscar Cooke

Land Below The Wind (Reprint)
by Agnes Keith

Three Came Home (Reprint)
by Agnes Keith

A Sabah Gazetteer
by J. Tangah & K.M. Wong

In Brunei Forests: An Introduction to the Plant Life of Brunei Darussalam (Revised edition)
by K.M. Wong

A Walk through the Lowland Rainforest of Sabah
by Elaine J.F. Campbell

Manual latihan pemuliharaan dan penyelidikan hidupan liar di lapangan
   by Alan Rabinowitz (*Translated by* Maryati Mohamed)

Enchanted Gardens of Kinabalu: A Borneo Diary
   by Susan M. Phillipps

The Theory and Application of a Systems Approach to Silvicultural Decision-making
   by Michael Kleine

Dipsim: A Dipterocarp Forest Growth Simulation Model for Sabah
   by Robert Ong & Michael Kleine

Kadazan Dusun–Malay–English Dictionary
   edited by Rita Lasimbang *et al.*

An Introduction to the Traditional Costumes of Sabah
   edited by Rita Lasimbang & Stella Moo-Tan

Traditional Stone and Wood Monuments of Sabah
   by Peter R. Phelan